Biology for the Individual BOOK 6

Working with Very Small Things

Donald Reid

Assistant Director (Schools),
The Health Education Council, London
formerly Head of Science Department,
Thomas Bennett School, Crawley, West Sussex

Philip Booth

Head of Science and Mathematics,
Castell Alun High School, Hope, Clwyd
formerly Head of Biology Department,
Thomas Bennett School, Crawley, West Sussex

 **Heinemann Educational Books**
London

Heinemann Educational Books Ltd
22 Bedford Square, London WC1B 3HH

London Edinburgh Melbourne Auckland
Singapore Kuala Lumpur New Delhi
Ibadan Nairobi Johannesburg
Exeter (NH) Kingston

ISBN 0 435 59760 4

First published 1974
Reprinted 1975, 1978, 1980, 1982, 1987

Filmset by Keyspools Limited, Golborne, Lancashire
Printed and bound in Great Britain by
BAS Printers Ltd, Over Wallop, Hampshire

Acknowledgements
We should like to thank—
—International Computers Ltd for their generous financial support

—Mr Maurice Elwell and his colleagues in the Nuffield Combined Science team for their assistance in the preparation of this book

—Our colleagues in trial schools, and especially the science staff at Brockworth School, Glos., for their constructive criticism of the draft version

The series was originally written as part of a teaching project initiated by the Nuffield Resources for Learning Project.

Acknowledgements for photographs—
Front cover: Ink cap fungus, H. R. Allen (N.H.P.A.)
Back cover: Paramecium, N.H.P.A.
Page 2: Fly agaric, and page 10, birch bracket fungus: Gordon F. Woods, F.R.E.S., F.R.P.S., (N.H.P.A.)
Page 27: Mount Everest Foundation. Picture shows Hilary and Tenzing.
Page 34: Mother and baby, John Curtis, Crawley;
Page 41: Louis Pasteur, Mansell Collection.
Page 54: Fleming's plate, St Mary's Hospital Medical School.

Contents

1 Looking at moulds 2
2 Looking at very small animals 4
3 Looking at bacteria 6
4 The four kinds of microbe 9
5 How do microbes affect our lives? 10
6 How small are small things? 12
7 What is the quickest way to make yeast ferment? 16
8 Looking at yeast 19
9 Making wine 20
10 Making bread 22
11 How to prevent soup from going bad 26
12 Where are microbes found? 28
13 Are there microbes in the soil? 30
14 Where else can microbes be found? 35
15 Are there microbes in your food? (pasteurization) 40
16 How fast can bacteria grow? 48
17 How to kill microbes in the home (disinfectants) 52
18 How to kill bacteria inside you (penicillin) 54
19 Where do microbes come from? 58

To the teacher
This book contains a collection of worksheets and short programmed
sequences, written to assist the teaching of elementary microbiology.
It is not intended to be a complete course, and we assume that
teachers will pick and choose from the contents.

Please consult the **Teachers' Guide to Books 1–7** in this series for an apparatus
list and further advice. It is also **essential** to consult the DES pamphlet **The Use of
Microorganisms in Schools**, Education Pamphlet 61 (HMSO 1977). Your
attention is especially drawn to the **safety advice** in paragraphs 21–45.

This is one of a series of books covering selected topics from
the Nuffield Biology and Combined Science courses. They are
suitable for pupils in all types of secondary school and
have been validated in a total of 80 different schools.

1 Looking at moulds

This book is about the very small living things called **microbes.**
They include moulds, bacteria, and viruses, and they affect our lives in many different ways.

Most microbes can only be seen with a microscope, but the moulds which grow on stale bread or rotting fruit can be seen by anyone.

Moulds, together with mushrooms, toadstools, and yeasts, are known as **fungi.**

Different kinds of fungi.

fly agaric fungus

shaggy ink cap fungus

moulds growing on fruit

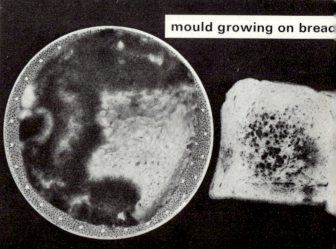

mould growing on bread

Most fungi feed on dead material such as old wood, fallen leaves or damp bread. Each fungus consists of a mass of thin threads which grow across its food and take in nourishment from it.

Fungi reproduce by making spores which are carried away by the wind. If a spore lands on a piece of bread left in your kitchen, it may begin to grow and produce threads.

Mould growing on bread,
magnified by 75 times (× 75)

Mushrooms and toadstools are simply large, complicated spore cases. They are formed from the threads like this—

How a toadstool is formed

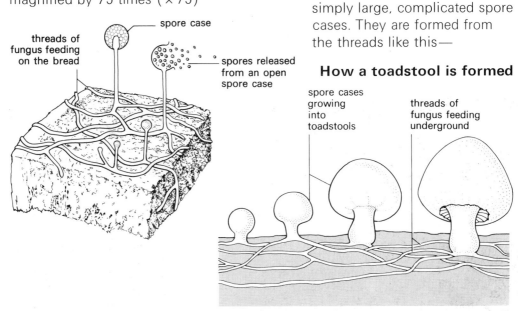

spore case

threads of fungus feeding on the bread

spores released from an open spore case

spore cases growing into toadstools

threads of fungus feeding underground

1 Now collect a small piece of food with mould on it, and a hand lens.

2 Look at your mould with the lens, and try to find the threads and spore cases.

3 Now collect a microscope, a mounted needle, a clean slide and a cover glass.

4 Put a little mould onto the slide and add a drop of water.

5 Now use the mounted needle to lower a cover glass carefully onto the mould.

6 **Look at your mould with the low power lens and make a drawing of it to show threads, spore cases, and spores.**

Now answer these questions—
  (a) What is your mould feeding on?
  (b) How do moulds obtain nourishment from their food?
  (c) How do moulds reproduce?
  (d) Most moulds release millions of spores at a time. Why do you think so many spores are made?

2 Looking at very small animals

Some microbes move about rapidly in water, where they feed on smaller living things such as bacteria. These fast moving microbes are mostly small animals called **Protozoa.**

To see Protozoa for yourself, follow these instructions—

1 Collect a mounted needle, a clean slide and cover glass, a syringe or dropper, and a microscope.

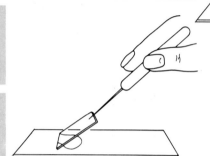

2 Use the dropper to collect one drop of liquid from a Protozoa culture, and place it on your slide.

3 Use the mounted needle to add a cover glass, as shown—

4 Use the low power lens to find small moving animals and then change to high power.

5 Find one of the largest Protozoa you can see. Look at the pictures on page 5 and try to find out its name. Use a reference book if you need more help.

6 **Find an animal which is moving more slowly than the rest and try to make a drawing of it. Label any parts you can see—** use the names on page 5 to help you, but do not copy your drawing from page 5. Draw only what you can see through your microscope.

Now answer as many of these questions as you can— use the information on page 5 if you need help.

(a) How many different kinds of Protozoa can you find in your drop of water?

(b) What are the largest Protozoa doing while you are watching them? (Moving slowly? Staying in one place? Moving rapidly?)

(c) How do the largest Protozoa move about? (By using legs? By using tiny hairs?)

(d) What do you think they are feeding on? How do they collect their food?

(e) Write down anything else you notice about them.

Different kinds of Protozoa, all magnified by about 500 times.

1 Paramecium (pronounced 'Para-meesium')

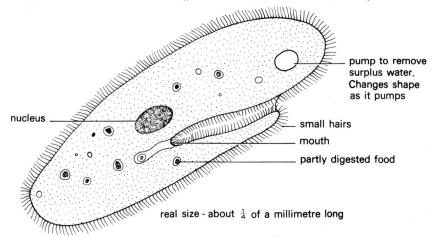

real size - about ¼ of a millimetre long

Paramecium uses its small hairs to move about in the water. It feeds on smaller microbes, such as bacteria. They are swept into its mouth by the small hairs.

2 Stylonychia (pronounced 'Style-o-nykia')

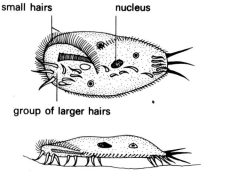

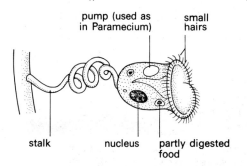

Stylonychia 'walking'

Stylonychia moves and feeds like Paramecium, but also seems to 'walk' using its larger hairs.

3 Colpidium

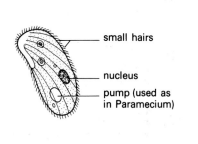

Colpidium moves and feeds in the same way as Paramecium.

4 Vorticella (pronounced 'Vorti-sella')

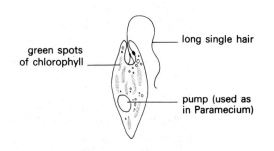

Vorticella lives mostly fastened by its stalk, but sometimes swims about. The hairs are used to catch microbes. Look for a small 'whirlpool' in front of Vorticella, caused by the small hairs.

5 Euglena (pronounced 'U-gleen-ah')

Euglena moves about like an animal, using its long hair to pull it through the water. However, it feeds like a plant, because it makes its own food with the help of the green spots.

3   Looking at bacteria

Introduction

Many kinds of bacteria are found in grass. If you boil some grass, you will kill all the bacteria in the grass, except for a few which form tough, resistant cells called **spores.**

If you then leave the boiled grass water for a few days in a warm place, the spores will turn into active bacteria again. They will soon multiply rapidly, feeding on chemicals in the water produced by boiling the grass. By boiling the water, you will also have destroyed small animals, such as Protozoa, which feed on the bacteria.

Instructions for growing your own bacteria.

1 Collect a beaker, a pair of scissors, a handful of fresh or dried grass, and heating equipment.

2 Put about 100 cubic centimetres of water into your beaker and then cut up the grass so that it fits inside the beaker.

3 Heat the water until it boils, and then turn down the flame and **boil very gently** for 15 minutes.

4 Collect a clean test tube, a piece of cotton wool, and a label.

5 Fill the test tube about three-quarters full with the 'boiled grass water'.

6 Plug the tube with cotton wool, label it, and leave it in a warm place for 2–3 days.

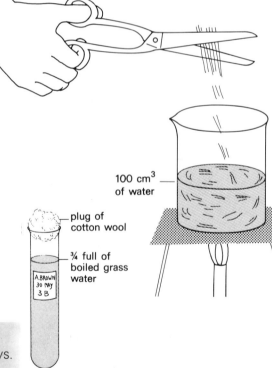

100 cm³ of water

plug of cotton wool

¾ full of boiled grass water

A. BROWN 30 May 3 B

Now write a short account of the experiment and then answer these questions— *(hint—read the introduction again)*
(a) Why is it necessary to boil the water?
(b) How can some kinds of bacteria survive boiling?
(c) Why is it necessary to plug the tube with cotton wool?

Looking at your own bacteria

1 Collect a microscope, two clean slides, two clean cover glasses, a mounted needle, a dropper, and your test tube of boiled grass water.

2 Use the dropper to collect one drop of liquid from the surface of the water in the tube.

take your drop from the surface of the water

3 Put the drop of liquid on the slide and cover it with a clean cover-glass.

grass bacteria × 200

single dots - each dot is one bacterium

chain of cells

4 Use the low power lens to look for tiny dots or chains of cells. When you have found some cells, change to high power. Your slide may look like this⟶

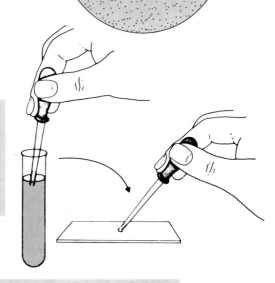

5 **To make your bacteria show up more clearly,** you can stain them with a blue dye.
To start with, collect a fresh drop of water from the surface of your test tube, and place this on your slide.

6 Leave the water on the slide to dry. **To speed up drying** leave the slide over an infra-red lamp, or near a convector heater, or a bunsen flame.

7 When the slide is dry, add a drop of methylene blue dye. Leave the dye in place for two minutes, and then carry on to page 8⟶

METHYLENE BLUE

8 After two minutes, wash off the dye with a few drops of water, over a sink.

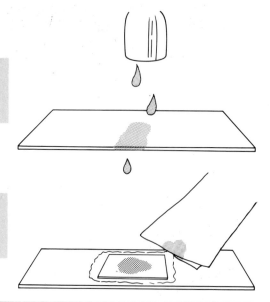

9 Add a cover glass and blot up any extra water with a paper towel.

10 Look at your slide with a microscope and make a drawing of two or three bacteria as seen under the high power lens.

Write down the magnification, for example '× 200'.

Answer these questions—

(a) Are your bacteria mostly found in clumps or mostly as separate, single cells?

(b) Are they all the same shape?

(c) Are they all the same size?

(d) If a graticule is available, try to measure the length of one of your bacteria in micrometres.

Further work

If microslides are available, collect a microviewer and the card called *'Harmful bacteria, set 20'*.

Follow the instructions to make sure that you insert the microslide correctly. Then answer these questions by looking at the slides and reading the notes on the card—

1 Why do typhoid bacteria have long hairs, called flagellae, on them?
2 Which bacterium was the main cause of death in the American Civil War?
3 Why are pneumonia bacteria difficult to kill inside the body?
4 How are invading bacteria, such as streptococci, killed in the blood?
5 Why is the botulism bacterium so deadly to man?

4 The four kinds of microbe

1 Fungi

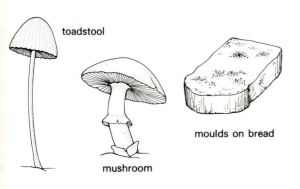

toadstool

moulds on bread

mushroom

Fungi feed mostly on dead plants or animals and so cause decay.

2 Protozoa. Minute animals. Each consists of one cell only.

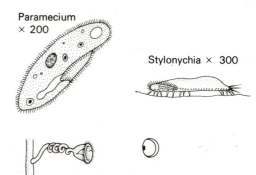

Paramecium × 200

Stylonychia × 300

Vorticella × 40

the protozoan which causes malaria × 5 000

Protozoa live mostly in water, feeding on bacteria. However, a few Protozoa cause disease.

3 Bacteria. Small cells, with many different shapes.

bacteria which feed on dead plants × 3 000

bacteria which feed on dead animals × 3 000

typhoid bacteria × 3 000

Most bacteria help to cause decay, but some cause disease.

4 Viruses. The smallest living things.

mumps virus × 15 000

smallpox virus × 15 000

polio virus × 60 000

a virus which attacks bacteria × 50 000

All viruses live inside living animals or plants. They always cause disease.

1 In the drawing of Paramecium, what does '× 200' mean?
2 Which are the smallest microbes shown?
3 Which kind of microbe is deliberately encouraged to grow in sewage works, because they feed on bacteria in the water?
4 Suppose all the bacteria and fungi in the world suddenly died out. What would happen to all the bodies of animals and plants which died from then on?

5 How do microbes affect our lives?

1 What made all these people ill?

She has measles

He has scarlet fever

He has caught a cold

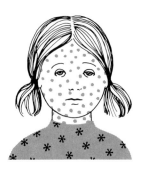

2 What caused these plant diseases?

black spots on rose leaves

the death of a birch tree

3 These cheeses were all made from milk.
What turns the milk into cheese?

4 Now go and look at the exhibition in your room and answer any questions you may find there.

Then answer the questions on this page.

5 Make a list of things which are made with the help of microbes.

6 Make a list of all the ways in which microbes can be harmful to us.

7 Suggest ways of preventing microbes from doing harm to us.

Further work or homework

8 **Find out how sewage is purified.** Give a diagram if you can.

9 Find out how microbes are used to make compost.
What does a gardener use compost for?

10 Find out as much as you can about how different cheeses are made.

11 Make up a large diagram to show all the uses of microbes you can think of.

12 Make up a large diagram to show all the harm which microbes can do to us.

6 How small are small things?

1 Since most microbes are too small to be seen, they are very difficult to measure.

For example, the small animal called Paramecium is one of the largest common Protozoa, but it can only just be seen without a microscope—

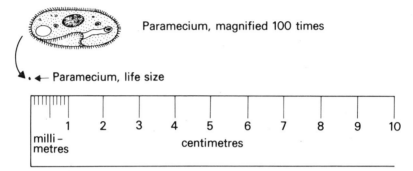

Paramecium, magnified 100 times

← Paramecium, life size

Paramecium is about one-quarter of a millimetre long, and so we cannot easily measure it in millimetres. Instead, we use **MICROMETRES.**

1 000 micrometres = 1 millimetre

Here is 1 millimetre, magnified by 100 times—

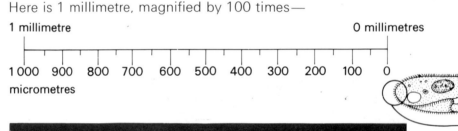

Paramecium also magnified 100 times (× 100)

How long is Paramecium in micrometres, roughly?

2 Paramecium is about 250 micrometres long.

Although bacteria are much smaller than animals like Paramecium, they can still be measured in micrometres—

Part of a Paramecium, × 4 000

Tuberculosis bacteria, × 4 000

Pneumonia bacteria, × 4 000

How long is the smallest bacterium shown above?

3

Although we can measure Protozoa and bacteria in micrometres, viruses are too small even for micrometres—

a group of viruses
attacking a
bacterium

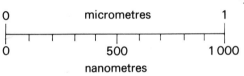

virus, magnified 60 000 times
(× 60 000)

bacterium, × 60 000

1 micrometre

So we measure viruses in **NANOMETRES.**

1 000 nanometres = 1 micrometre
1 000 micrometres = 1 millimetre

The virus which causes flu, × 60 000

Polio virus, × 60 000

Mumps virus, × 60 000

0	micrometres	1

0   500   1 000
nanometres

What is the diameter of the largest virus shown above?

4

We can also measure very large molecules in nanometres—

Polio virus, × 600 000

*A molecule of the protein
in egg white, × 600 000*

0   20   50
nanometres

So the smallest viruses are not much larger than the largest molecules.

But atoms are very much smaller still, and we measure them in
PICOMETRES (pronounced *'pyko-metres'*)
1 000 picometres = 1 nanometre or 1 picometre = 10^{-9} millimetres
1 000 nanometres = 1 micrometre or 1 nanometre = 10^{-6} millimetres
1 000 micrometres = 1 millimetre or 1 micrometre = 10^{-3} millimetres

What is the diameter of this hydrogen atom?

*Hydrogen atom,
magnified by twenty million times*

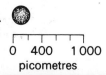

0   400   1 000
picometres

Summary of the sizes of very small things

5

The largest one celled microbes are the small animals called Protozoa—

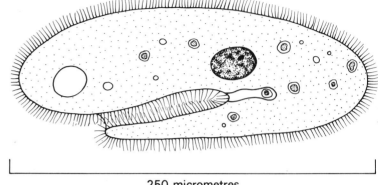

Paramecium, ×400

250 micrometres

The next largest microbes are the bacteria—

typhoid bacteria ×400

6

But bacteria are much larger than viruses or large molecules—

typhoid bacterium, ×60 000

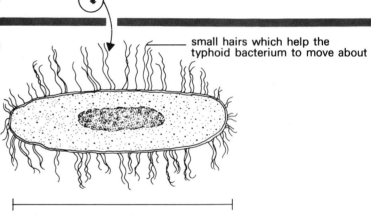

small hairs which help the typhoid bacterium to move about

1 micrometre = 1 000 nanometres

mumps virus ×60 000

225 nanometres

yellow fever virus ×60 000

40 nanometres across

a very large molecule (the protein from egg white) ×60 000

10 nanometres across

7

And protein molecules are very much larger than atoms—

10 000 picometres   (10 nanometres)

*a very large
protein molecule
from egg white,
magnified 10 million
times.*

*a hydrogen atom,
also magnified
10 million times* 200 picometres across

Test

1 **(a) Write down the name of the smallest object on the list below—**
a mumps virus
a large molecule
Paramecium
a hydrogen atom
a typhoid bacterium

**(b) Now write down the name of the next smallest object, and
then the next smallest, and so on.** The last name will be the
name of the largest object.

2 How many picometres make one millimetre?

**3 How many hydrogen atoms,
each 200 picometres across,
would fit across the sharp**   sharp point of a pin
point of this pin?   100 micrometres across

Further work

**4 Find out if there are any smaller units than picometres.
What could be measured with these units?**

7 What is the quickest way to make yeast ferment?

1 Collect a beaker, and half fill it with lukewarm water.
Then set it up as shown and try to keep the temperature of the water as close to 45°C as possible.
Beware—DO NOT OVERHEAT.

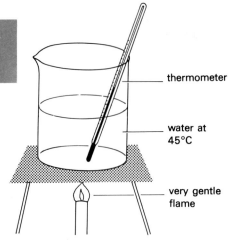

thermometer

water at 45°C

very gentle flame

2 Set up one test tube like this—

label the tube A

1 add **yeast** to about ½ cm deep

2 add **sugar** to about ½ cm deep

3 add 2 cubic centimetres of **hot water at 70°C**

3 When you have finished adding the yeast, sugar, and hot water, the test tube should be only **one-third** full.
If it is too full, throw away some of the water.

4 **Shake your tube** very thoroughly for 10 seconds.
Then turn it upside down and shake again for 10 seconds.

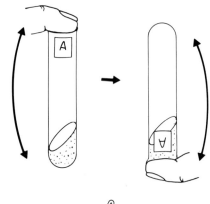

5 Place the tube in the water bath.
Check that the temperature of the water is about 45°C.

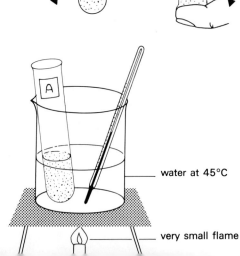

water at 45°C

very small flame

6 Now set up three more test tubes like these—

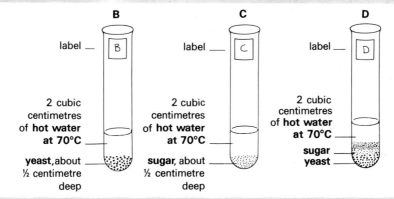

	B	C	D
label	B	C	D

B — 2 cubic centimetres of **hot water** at 70°C — yeast, about ½ centimetre deep

C — 2 cubic centimetres of **hot water** at 70°C — sugar, about ½ centimetre deep

D — 2 cubic centimetres of **hot water** at 70°C — sugar, yeast

7 Make sure you have labelled each tube as shown.

8 **Shake each tube thoroughly** for 5 seconds, and then turn upside down and shake again for 5 seconds.

9 Place B and C in the beaker with A.
Leave D on the bench.

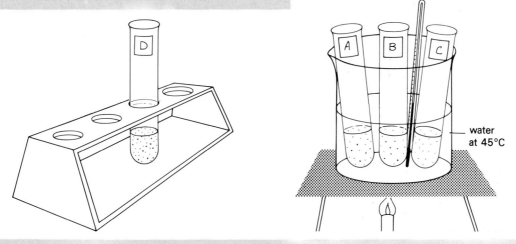

water at 45°C

10 If the liquid in the tubes starts to froth, then a gas is being given off.
Test the gas like this—

1 Put a **glowing splint** into the tubes

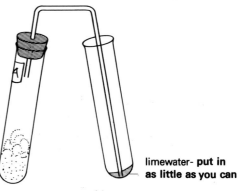

2 Fit up a delivery tube and pass the gas through **limewater** for at least ten minutes

limewater- **put in as little as you can**

Now turn over——→

11 Write down your results in a table like this—

What is the quickest way to make yeast ferment?						
Tube	Contents	Temperature	Was there a steady stream of bubbles up the tube?	Did the splint flare up or go out?	Did the limewater turn chalky?	Which gas was given off?
A	Yeast, sugar, and water	45°C				
B	Yeast and water only	45°C				
C	Sugar and water only	45°C				
D	Yeast, sugar, and water	15–20°C				

Write up the experiment in the usual way, and then answer these questions—

(a) In which tube was the most gas produced?

(b) Why was it important to set up tubes B and C?

(c) Why was tube D set up?

(d) Why were you told to use as little limewater as possible?

(e) Recipes for home made wine usually suggest mixing yeast, sugar, fruit juice, and water, and then leaving the mixture to ferment in a warm place.

Why is sugar needed to make wine?

8 Looking at yeast

1 **Collect**—a clean microscope slide, a cover glass, a dropper, a mounted needle, and a microscope.

2 Use the dropper to place a drop of yeast culture on your slide.

3 Now add a drop of **blue dye** to make the yeast show up more clearly.

4 Use the mounted needle to place a **cover glass** on the slide.

5 Now look at the yeast with your microscope, and then answer these questions by looking at **your** yeast cells—

1 **What do your yeast cells look like, down the microscope?**
(square? round? egg-shaped?)

2 **Are all your yeast cells the same size?**

3 **Yeast cells can grow by 'budding', as you can see in the photograph—**
Are any of your cells budding?

4 **Draw two or three of your yeast cells, as seen under the high power lens.**
Draw them as large as you can.

5 **Put a scale on your drawing, for example 'x 100', to show the magnification you used.**

6 **How large do you think they really are?**
Use a graticule, if one is available.

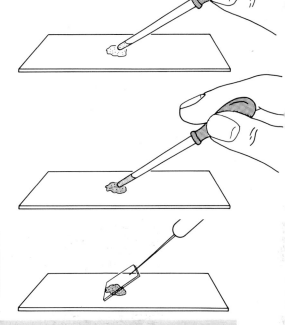

Yeast cells, magnified 3 500 times.

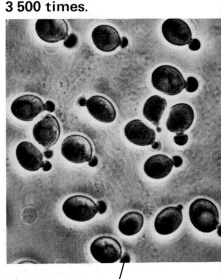

small yeast
off from larger ones

9 Making wine

1 Collect four test tubes, a test tube rack, a test tube holder, and a bunsen burner.

2 Set up **three** tubes like this—

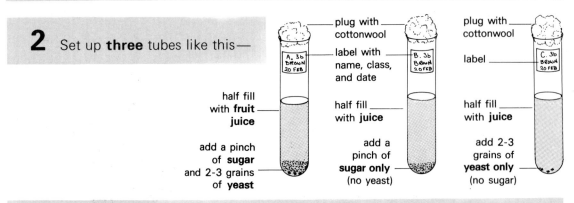

plug with cottonwool

label with name, class, and date

half fill with **fruit juice**

add a pinch of **sugar** and 2-3 grains of **yeast**

plug with cottonwool

label

half fill with **juice**

add a pinch of **sugar only** (no yeast)

plug with cottonwool

label

half fill with **juice**

add 2-3 grains of **yeast only** (no sugar)

3 Now set up the fourth tube exactly like A, but then **boil it gently** for two minutes.   Label the fourth tube **'D'**.

4 Copy out this results table—

Making wine		Date set up	
Tube	Contents	Result after a few days	
		Were any bubbles seen?	Did the tube smell differently?
A	Yeast, fruit juice, sugar		
B	Fruit juice and sugar only		
C	Yeast and fruit juice only		
D	Yeast, fruit juice, and sugar—boiled		

5 Now write up the experiment in your usual way and then answer the questions on page 21.

Questions on wine-making

1 (a) Why was it necessary to plug your tubes with cotton wool?

(b) You could also have plugged your tubes with a solid rubber bung or cork.
This would not have been a good idea.
Why not?

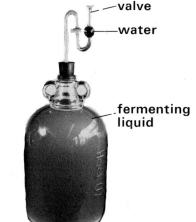

2 Home made wine can be made from fresh **or** from canned fruit juice.
If you use canned juice, you must first make sure that there is no preservative in the juice.
Why?

3 Most books on wine making recommend that the mixture be left to ferment in a large jar, fitted with a valve like this—

(a) Why is this valve used?

(b) How does it work?

4 (a) Which gas is given off during fermentation?

(b) How could you test it to find out?

5 Anyone who makes home made beer or wine, will tell you that a hydrometer like these, is very useful—

(a) What does a hydrometer measure when it is placed in any liquid?

(b) Why is a hydrometer so useful to someone making home made beer or wine?

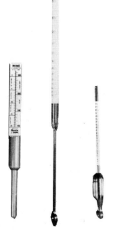

10 Making bread

First recipe—making bread with yeast

1 Wash your hands with soap and water.

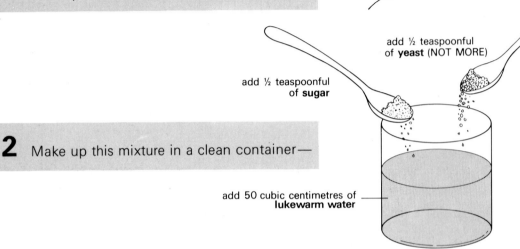

add ½ teaspoonful of **yeast** (NOT MORE)

add ½ teaspoonful of **sugar**

2 Make up this mixture in a clean container—

add 50 cubic centimetres of **lukewarm water**

3 **Stir the mixture and leave for about 10 minutes,** until you can see bubbles appearing. **Meanwhile, carry on.**

4 **Collect a sheet of newspaper** and spread it on your bench. Make sure you do all the mixing of dough on the paper, to prevent mess.

5 Collect a clean mixing bowl and add flour and salt, as shown—

add 2 heaped tablespoons of **flour**

add 1 pinch of **salt**

6 **Now add your yeast and water mixture.**

7 Stir the mixture of flour, sugar, and yeast with a fork or spoon, and then 'knead' it to make dough as shown—

How to knead dough

1 Grasp the edges of the dough and pull into the centre

2 Push downwards with the knuckles

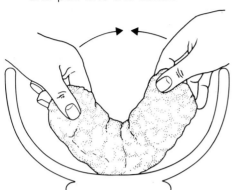
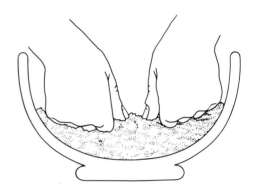

8 Continue kneading for several minutes until the dough leaves the fingers cleanly.

***If the dough is wet and sticky*—add a little more flour.**

***If the dough is dry and powdery*—add a little water.**

9 When your dough feels firm to the touch and leaves your fingers cleanly, place it in a small container, as shown—

Cover and label the container.

paper
towel
cover

label

A. BROWN
With yeast

10 Leave the container in a warm place for at least 30 minutes. Meanwhile, carry on.

11 Collect a paper towel and clean out your mixing bowl. **Do not wash the dough into a sink**—you may block the sink if you do. Place the dirty paper towel in a bin.

12 When your bowl is completely clean, try the second recipe on the next page——→

Second recipe—making bread without yeast

1 Add flour, sugar, and salt to a clean mixing bowl—

add two heaped tablespoons
of **flour**

add a pinch of **salt**

add half a
teaspoon of
sugar

do NOT add yeast

2 Add 50 cubic centimetres of water, and knead the dough until it feels firm but neither dry nor sticky.

3 Then place your dough in a second small container, cover it, and label it—

A BROWN
No yeast

4 Clean out your bowl with a paper towel as before. **Do not** allow any dough to fall into the sinks.

Now write up the experiment in your usual way.

Baking the dough

1 **Find the lump of dough made without yeast.** Take it out of its container and place it on a greased baking tray, in one of the numbered squares marked 'no yeast'.

2 Write down the number of the square and then **find your lump of dough made with yeast.** Place it on a numbered square marked 'yeast'. Write down the number of the square. **Are your two lumps still the same size?**

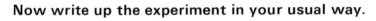

3 Bake at 225°C (gas Regulo 8) until brown. This will take between 30 and 60 minutes.

Questions on 'Making bread'

1 Why is it important to wash your hands before you begin?

2 Which lump of dough rose the most?

3 Dough rises when a gas is produced inside it.
   Where did this gas come from, in your dough?

4 (a) **Which gas do you think this was?**
   (b) **How could you test to find out which gas was present?**

5 (a) **Did the dough go on rising in the oven?**
   (b) **What do you think happens to the yeast in the oven?**

6 There is a story in the Bible about **'unleavened'** bread.
   To leaven means 'to lift'.
   How do you think unleavened bread is made?

unleavened bread

leavened bread

7 **What is the difference between plain flour and self-raising flour?**

8 **What makes cake mixture rise?**

11 How to prevent soup from going bad

1 Collect three test tubes, a container to hold the tubes, a funnel, and two small pieces of cotton wool.

2 Fill each tube about **half full** with the thin soup provided.

3 **Collect a felt tip pen** and label the tubes with your initials and the letters A, B, C, like this—

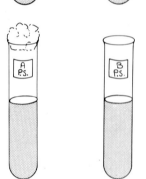

your initials

4 Roll the cotton wool to make two tight plugs for tubes A and C.

Leave B open to the air.

5 Collect two pieces of cooking foil and two elastic bands. Set up tubes A and C as shown—

The foil will keep the cotton wool dry during sterilizing in the cooker.

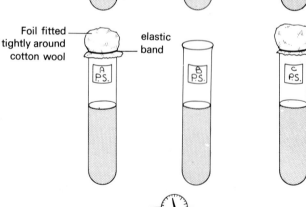

Foil fitted tightly around cotton wool

elastic band

6 Now take B and C to be sterilized in a pressure cooker at 120°C for 15 minutes.
Leave A unsterilized.

Meanwhile, read on——→

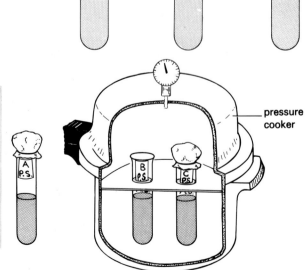

pressure cooker

7 Copy out this results table—

Tube	Treatment	Results Was the soup clear or cloudy? Did moulds grow inside the tube?			
		At the start	After 1 week	After 2 weeks	After 3 weeks
A	covered, not sterilized.				
B	open, sterilized.				
C	covered, sterilized.				

How to prevent soup from going bad.   Date started:

Questions to be answered while the tubes are in the cooker—

1 Write up the experiment in your usual way.

2 At what temperature does water usually boil?

3 What is the temperature of the boiling water in your pressure cooker?

4 Why is it better to sterilize soup in a pressure cooker instead of boiling it in a beaker?

5 Climbers working at 8 000 metres (25 000 feet) on Mount Everest, cannot cook eggs quickly unless they use a pressure cooker.

Why do they need a pressure cooker?

6 Suppose tube C had been closed with a rubber bung.
(a) What would have happened when the tube was inside the cooker?

(b) Give a reason for your answer to question 6 (a).

12   Where are microbes found?

1 Are there microbes in the air?
　　　　—in the soil?
　　　　—on your hands?

You could find out by looking for them, but you are likely to find this very difficult, even with a powerful microscope.

Why are microbes very difficult to see?

2 The easiest way to look for microbes is to use dishes of nourishing jelly.
If a microbe falls onto this jelly, it will grow and multiply.
After a few days, a large spot containing millions of microbes can be seen—

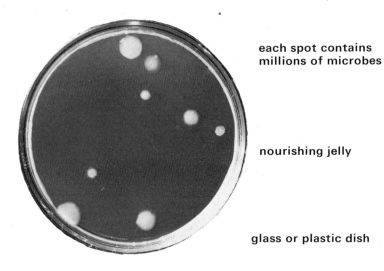

each spot contains
millions of microbes

nourishing jelly

glass or plastic dish

If each spot was formed from one microbe, how many microbes fell onto this plate originally?

3

These dishes of jelly can be used
to find out the answers to questions
such as—

'Are there any microbes in soil?'

For example—

1 Suppose this is a sterilized dish
containing sterilized jelly.

2 The lid is quickly opened
and a little soil,
mixed with purified water,
is added.

**soil mixed
with water**

3 The lid is replaced, and the
dish is kept for 2—3 days
in a warm place.

By then, microbes have grown
on the jelly.

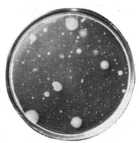

These microbes **may** have come from
the soil that was added to the jelly.
But you can't be sure.

Where else might they have come from?

4

Microbes might have come from—

● the air, when the dish was open

● the syringe, used to add the soil.

They may also have come from the dish itself, the jelly, or the water, if
these have not been perfectly sterilized.

**How could you check to find out whether any of the microbes
came from the air, the water, or the jelly, etc?**

13 Are there microbes in the soil?

To find out, you can make up two jelly dishes yourself, and then add soil to one of them. The other dish can be used for a comparison.

1 Set up a beaker of water, **half** full of hot water.

Leave the water to warm up until it is boiling. Meanwhile, carry on.

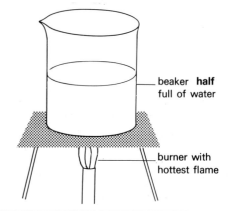

beaker **half** full of water

burner with hottest flame

2 Wash your hands thoroughly.

3 Collect— 2 sterile, empty dishes, with lids   2 sterile syringes
2 sterile screw cap bottles,   2 sterile paper towels
each containing sterile jelly   labels or felt tip pens
2 spring clamps, to hold the bottles

4 Keep the sterile dishes **upside down** to prevent the lids from coming off by accident.
Leave them on the sterile paper towels whenever possible.

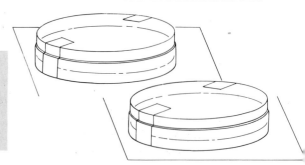

5 **Loosen the top of each bottle slightly.** This will make it easier to undo them later when they are hot.

6 **Fix a spring clamp** to each bottle so that the bottles can be easily lifted when hot.

7 **Place both bottles in the water bath** and leave them until the jelly melts. This will take about 5–10 minutes.

Meanwhile, carry on to page 31 ⟶

8 Collect a sterile beaker, and add to it about 100 cubic centimetres of purified water and one teaspoonful of soil.

teaspoonful of soil

100 cm³ of purified water

9 Stir the mixture, and then leave it to allow the larger grains of soil to settle on the bottom.

10 When the jelly in the bottles has melted, take one of the bottles out of the water bath.

Leave it on a paper towel, on the bench, to cool for a few minutes—**but do not let it set again.**

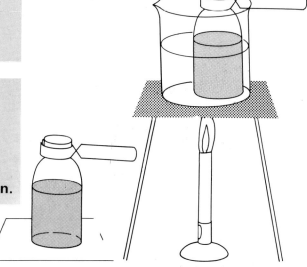

11 Meanwhile, draw up 1 cubic centimetre of your soil and water mixture into a syringe. **Place the tip of the syringe only just below the surface of the water,** to avoid sucking up large grains of soil.

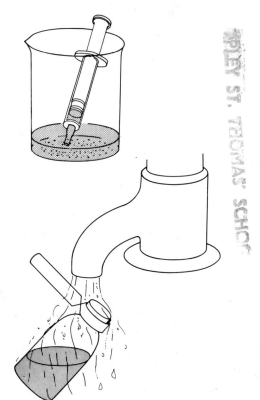

12 Now cool your bottle quickly under a cold tap—do not allow the jelly to set.

13 Remove the top of the bottle— use a paper towel if the top is too hot.

Now turn over ⟶

14 Add your 1 cubic centimetre of soil and water to the melted jelly in the bottle.

15 Pass the neck of the bottle through a bunsen burner flame, several times, quickly.

16 Turn one of your dishes over and pour the soil and jelly mixture into the dish. **Open the lid as little as possible and do not breathe on the jelly.**

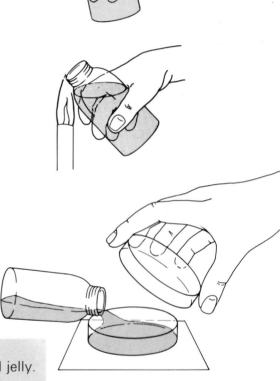

17 Rock the dish gently to mix the soil and jelly.

18 Now leave the dish flat on the bench until the jelly has set. This may take up to 15 minutes.

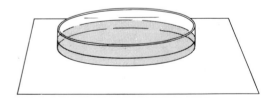

19 Meanwhile, label the lid as shown, and then carry on to the next page⟶

To set up the second plate—

1 Remove the remaining bottle from the water bath and leave it on the bench for a few minutes to cool.

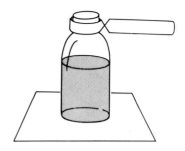

2 Use your second syringe to draw up 1 cubic centimetre of purified water.

3 **Now repeat stages 12–19 on pages 31–32,** so that you add the purified water to the second bottle of jelly.

4 Keep both dishes flat until the jelly has set. Label the dishes as shown.

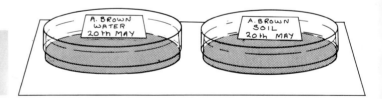

5 When the jelly has set, turn the dishes upside down. Stick the lid to the base of each dish, with sticky tape.

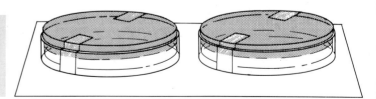

6 The dishes should now be left in a warm place for several days.
Now turn over————→

Questions on 'Are there microbes in the soil?'

1 Write a very short summary of the experiment, in about 10 lines. Use diagrams if possible.

2 The second dish, containing jelly and purified water only, is known as the 'control' dish. **Why was it necessary to set up a control dish?**

3 Suppose microbes grow on the control dish. **What will this prove?**

4 Why were each of these precautions necessary?
   (a) washing your hands before you began.
   (b) taking care not to breathe over an open dish.
   (c) keeping the dishes on a clean paper towel and not on the lab bench itself.
   (d) turning the dishes upside down after they had set.

5 The label on a can of evaporated milk has these instructions on it—

For baby feeding
● Always sterilize bottle, teat and equipment to be used in making the feed.

● Always clean top of can before opening, and scald with boiling water.

  (a) Why is it important to scald the top of the can with boiling water?
  (b) What may happen to the baby if its parents do not sterilize the feeding equipment before each feed?

6 After you have left your dishes in a warm place for 2–3 days, you will be looking at them to see if any microbes have grown on the jelly.
  (a) However, the dishes must not be opened. Why not?
  (b) When you have finished with your dishes, they should be placed in disinfectant. Why is it important to do this?

14 Where else can microbes be found?

1 You can find out by carrying out further experiments.
Remember these precautions, whichever experiment you choose—

- Wash your hands before you start (unless you choose experiment 3).
- Do not breathe on an open dish.
- Do not open the dishes after they have been incubated.

Now choose one of the experiments from this list and turn to the correct page for further advice—

A **Are there microbes in water?**   *See this page*

B **Are there microbes in the air?**   *See page 36*

C **Are there microbes in your hair?** *See page 36 (lower half of page)*

D **Are there microbes on your fingers or your teeth?** *See page 37*

E **Are there any other places where you would**   *See page 38*
   like to search for microbes?   *(lower half of page)*

2 Are there microbes in water?

1 Find out which different kinds of water are available. Choose one kind for your experiment.

2 Now design your own experiment to find out how many microbes are present **in 1 cubic centimetre** of the water you have chosen.

3 Don't forget—you must include a **control dish.**

      —you must **label** all your dishes and stick the lids to the bases.

4 Write up your experiment in the usual way, and explain why a control experiment was needed.

5 Turn to page **39** and answer question 1.

3 Are there microbes in the air?

1 You can find out by leaving a sterile dish of jelly with the lid off, for 30 minutes or more.

If enough dishes are available, you could do this in different places in the room, or even in different rooms.

2 Don't forget to set up a control dish also.

3 While your dishes are open to the air, write a short account of your experiment, and explain why you needed a control dish.

4 If you still have time, turn to page 39, and answer question 1.

5 When you have finished, remember to label your dishes and to stick the lids to the bases.

4 Are there microbes in your hair?

1 You can find out by combing your hair over a sterile dish of jelly—

2 Don't forget to set up a control dish also.

3 Label your dishes and stick the lids to the bases.

4 Write a short account of your experiment and explain why you needed a control dish.

5 Turn to page 39 and answer question 1.

5 Are there microbes on your fingers or your teeth?

1 Make sure that one of you has dirty hands.

2 Collect a sterile dish of jelly, turn it upside down and mark the base as shown—

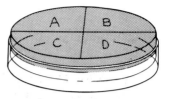

3 Collect a mounted needle and sterilize it by heating to red heat in a flame.

4 Now try to use different parts of the plate in different ways—for example—

Area A—touch the jelly gently from underneath with dirty fingers.

Area B—touch the jelly gently from underneath with clean fingers.

Area C—scrape your teeth with the sterile needle, and then gently stroke the jelly with the needle.

Area D—leave untouched.

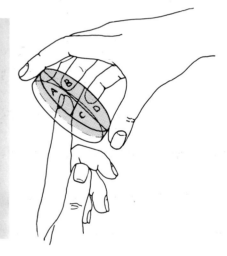

5 Now close the dish, seal it with sticky tape and label it with your name, class and the date.

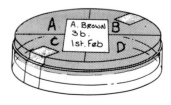

6 Turn over to page 38 ———→

7 Copy out this table of results—

Are there microbes on the fingers or teeth?		
Area of dish	How was it treated?	How many microbe colonies could be seen there, 2–3 days later?
A	touched with dirty fingers	
B	touched with clean fingers	
C	touched with teeth scrapings	
D	not touched	

8 Write up the experiment in your usual way.

9 Why is it important to leave one part of the dish untouched?

10 Now turn to page 39 and answer question 1.

6 Are there any other places where you would like to search for microbes?

1 Before you do anything, write down your answers to these questions—

2 What are you trying to find out?

3 How can you prove that your jelly dishes were sterile before you began? (Hint—use a control dish)

4 What equipment will you need?

5 Now see your teacher with your answers.

6 When your experiment is set up, write it up in the usual way, and then answer question 1 on page 39.

7 Questions on 'Where are microbes found?'

1 Collect the book called *'Biology for the Individual, Book 7—The war against disease'*, and turn to page 12, 'Death on the operating table'. **Read pages 12-17 and answer the questions on each page.**

Questions to be answered after your dishes have been incubated for 2–3 days—

2 **Write down the results of YOUR experiment.**

3 **Copy out this table, and fill it in from the CLASS results—**

Places where microbes were found	Places where microbes were not found

4 **Write down any reasons you can find to explain why microbes were present or absent in each of the places mentioned in the table.**

5 You are probably carrying several million bacteria about with you on your **skin.** These bacteria are usually harmless unless they get into a cut. Some skin bacteria are actually useful, because they kill harmful bacteria.

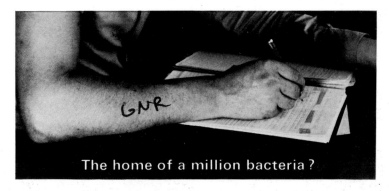

The home of a million bacteria?

Did the results of the class experiments suggest that there are a great many bacteria on the skin?

Homework suggestions

1 **Make a list of the places in your home where microbes can be found.**

2 **What steps can be taken to keep these places free from microbes?**

15 Are there microbes in your food?

No-one is ever likely to eat the fruit in the box shown above.

1

What has made the fruit go bad?

2

As you know, many foods will go bad unless they are protected from decay.

For example, milk soon begins to smell if it is left in a warm place for a day or two—

Milk can be kept fresh by boiling it.

Boiled milk is sold under the name *'sterilized milk'*.

bottle of sterilized milk

1 What does the word *'sterilized'* mean?

2 Are there any bacteria in sterilized milk?

3 Why is sterilized milk not as popular as other kinds of milk?

3

Sterilized milk is not very popular, because many people do not like its taste.

Another way to prevent milk from going bad is to keep it cool in a **refrigerator.**

But it would be very expensive to keep milk in a fridge all the way from the farm to your house—

A third way to prevent decay was discovered by the famous French scientist, **Louis Pasteur.**

He found that wine would keep better if it was heated to 63°C and then cooled quickly.

This killed most of the microbes in the wine, but did not alter its taste.

Louis Pasteur in 1852

This treatment is called **pasteurization** (pronounced 'pasteur-eye-zay-shun').

Many wines and beers are pasteurized nowadays to prevent microbes from growing in them after fermentation is complete.

Wine which is not pasteurized is likely to become sour, and may even turn into **vinegar.**

Why do you think this method of preserving drinks is called PASTEURization?

You can find out more about pasteurization and refrigeration by carrying out this experiment—

1 Collect a beaker (250 cm³), a thermometer, and heating equipment. Place about 150 cubic centimetres of warm water in the beaker.
Heat gently until the temperature reaches 63°C.

63°C

two thirds
full of
warm water

2 Collect three screw cap bottles and fill each to the neck with untreated milk.

Make sure that the tops are screwed on firmly.

milk

3 When the water in the beaker has reached 63°C (not more), place **one** of the bottles in the bath.

leave at
63°C for
30 minutes

4 Leave this bottle in the bath for 30 minutes. Meanwhile, carry on.

5 Keep the other two bottles on the bench. Label them B and C and add your name, class and the date.

B
A. BROWN
3b
20 JUNE

C
A. BROWN
3b
20 JUNE

6 Check that the water temperature is at 63°C and then carry on to page 43 ——→

7 Copy out this table—

Does pasteurization or refrigeration help to keep milk fresh?				
At the start		A few days later		
1 Bottle	2 How was it treated?	3 Did it smell fresh or sour?	4 Did it clot on boiling?	5 Was it fresh, slightly sour, or very sour?
A	pasteurized			
B	kept in a fridge			
C	not pasteurized or kept cool			

8 Check again that the temperature is 63°C, and then read pages 48–50 in this book, answering any questions you may find there—**but remember to turn back to this page when your bottle has been in the water for 30 minutes.**

9 When the 30 minutes is over, remove the bottle from the water and cool it in cold, running water. **Keep the bottle under cold water for at least one minute—time yourself!**

cool the bottle for at least one minute

10 Label the bottle as shown—

11 Leave bottle B in a fridge, and leave bottles A and C in a warm place, for several days.

Write up the experiment in your usual way, and then finish answering the questions from pages 48–50.

Testing your milk for freshness

1 Collect three large test tubes, a test tube rack, and a bunsen burner.

2 **Shake** each of your bottles of milk, very thoroughly.

3 Open each bottle and smell the milk, **but do not try to taste it.** Dangerous bacteria may be present.

Fill in column 3 of your results table.

4 To test for freshness more accurately, first pour a little milk from bottle A into a large test tube.

5 **Heat the milk with a very gentle flame,** until it boils. Remove the test tube from the flame as soon as the milk boils.

use a very gentle flame

6 Shake the tube gently, and look for small clots.

Sour milk clots easily on boiling, but fresh milk does not clot.

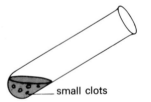

small clots

7 Fill in column 4 of your results table, and then keep the boiled milk in its tube, in the rack, for comparison with the other tubes.

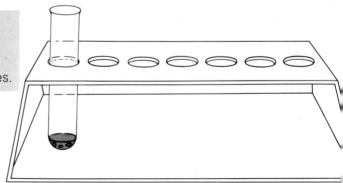

8 Repeat stages 4–7 with the milk from bottles B and C. **Do not throw away the milk in the bottles**—you will need it for the next experiment.

Growing the bacteria from fresh and sour milk

Does sour milk contain more bacteria than fresh milk?
To find out, follow these instructions—

1 Collect a chinagraph pencil, a wire loop, a bunsen burner, and a sterile dish containing a special jelly for growing the bacteria from milk.

2 Label the base of the dish with numbers and your name, and class, as shown—

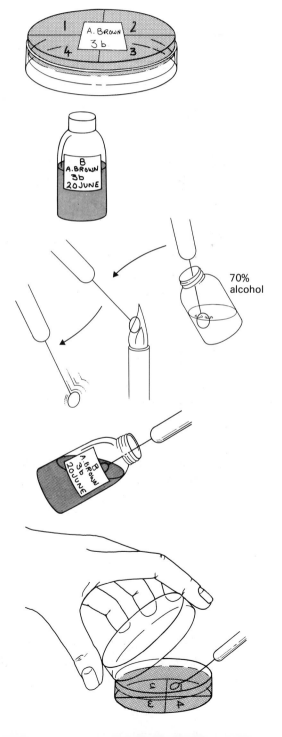

3 Collect a bottle of fresh milk from the previous experiment. **Use the freshest milk you can find.**

4 Dip your wire loop into a little 70% alcohol and drain it against the side of the alcohol bottle. Light your bunsen burner, and hold the wire loop in the flame until it is red hot.

Then cool it in the air.

5 Collect a small drop of fresh milk on the loop.

6 Turn your jelly plate over and streak the loop gently across the jelly in area 1. **Do not let the loop dig into the jelly.** Make sure that the loop touches the jelly only in area 1.

Carry on to page 46 ——→

7 Close the lid of the dish, and find a bottle of milk which is slightly sour.

8 Sterilize the loop again, and cool it in the air. Streak a drop of milk from the second bottle across area 2.

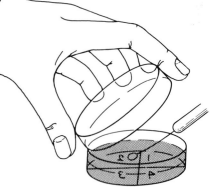

9 Close the lid of the dish and **collect a bottle of the sourest milk you can find.**

10 Sterilize the loop again, and cool it in the air. Streak a drop of milk from the bottle of sourest milk across area 3.

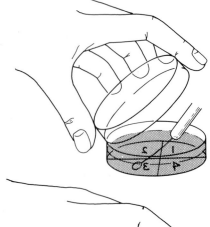

11 Close the lid of the dish. Sterilize the loop again, and cool it in the air.

Now streak the empty loop across area 4.

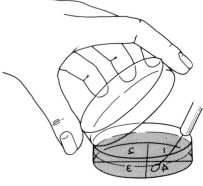

12 Seal up the dish with sticky tape. Make sure you have labelled the base with your name and class. Leave the dish in a warm place for 2–3 days.

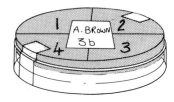

13 Now copy out this results table—

Does sour milk contain more bacteria than fresh milk ?		
Area of jelly	Type of milk used	Number of colonies of bacteria seen in each area
1	freshest	
2	slightly sour	
3	sourest	
4	none	

Write up the experiment in the usual way.

Questions on keeping milk fresh

1 In the first experiment on the pasteurizing of milk (on page 42), you were told to set up a bottle of milk, which was neither pasteurized nor kept cool (bottle C).
This was the **control** for the experiment. **Why was it important to set up a control bottle ?**

2 In the dish of jelly which you have just set up, area 4 was streaked with an empty loop. So area 4 was a control area.
Why was it important to have a control area which was not streaked with milk ?

3 From the class results, which seems the better way to preserve milk—
pasteurization or refrigeration ?

4 Why does milk stay fresh if it is kept cold ?

5 Pasteurization does not kill all the bacteria in the milk, so why is it carried out ?

6 Which famous disease used to spread very easily in milk which had not been pasteurized ?

16 How fast can bacteria grow?

1

If milk is not pasteurized or kept cool, it turns sour very quickly.
You can find out why by reading on—

Suppose there is just one bacterium in this bottle of milk.
The bottle is left standing in the sun at 8.30 a.m.

8.30

One bacterium at 8.30 a.m.
grows and divides—
to give **two bacteria** by 9.00 a.m.

9.00

Two bacteria grow and divide—
to form **four bacteria** by 9.30 a.m.

9.30

Four bacteria grow and divide—
to give **eight bacteria**
by 10.00 a.m.

10.00

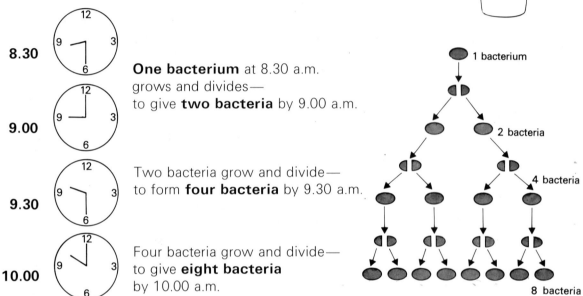

1 bacterium

2 bacteria

4 bacteria

8 bacteria

So during hot weather, bacteria can divide every 30 minutes.

How many bacteria will there be in the bottle at 10.30 a.m.?

2

There will be sixteen bacteria in the bottle at 10.30 a.m. because—

eight bacteria at 10.00 a.m.

—become

sixteen bacteria by 10.30 a.m.

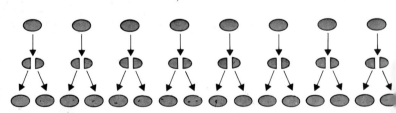

How many bacteria will there be in the bottle, sixty minutes later, at 11.30 a.m.?

3

After 10 hours, there could be over **1 million (10⁶)** bacteria in the milk.

After 24 hours, if there was enough food in the milk and enough space in the bottle, there would be over **1 billion (10¹⁴)** bacteria in the milk.

Obviously, this never happens, because growth slows down long before then, as you can see from this graph—

How one bacterium in a bottle of milk grows to 32 million in a few hours

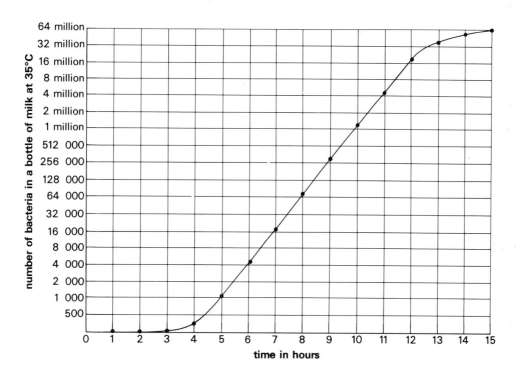

1 When did growth first start to slow down? Choose—
   (a) between 11 and 12 hours
   (b) between 12 and 13 hours
   (c) between 13 and 14 hours

2 Why do you think that the bacteria in a bottle of milk are likely to stop growing after a few hours?

3 The number of people in the world is also growing, though not as fast as the bacteria in the milk.
Our rate of growth will also slow down in the end.

Why is it impossible for the number of people in the world to continue increasing for ever?

We have been looking at the effects of keeping food in warm or cold conditions. This diagram is a summary of the effects of heat on bacteria—

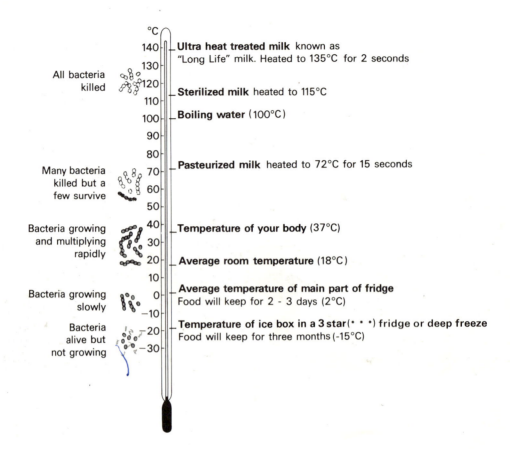

°C

140 — **Ultra heat treated milk** known as "Long Life" milk. Heated to 135°C for 2 seconds

130

All bacteria killed 120

110 — **Sterilized milk** heated to 115°C

100 — **Boiling water** (100°C)

90

80

70 — **Pasteurized milk** heated to 72°C for 15 seconds

Many bacteria killed but a few survive 60

50

Bacteria growing and multiplying rapidly 40 — **Temperature of your body** (37°C)

30

20 — **Average room temperature** (18°C)

10

Bacteria growing slowly 0 — **Average temperature of main part of fridge** Food will keep for 2 - 3 days (2°C)

−10

Bacteria alive but not growing −20 — **Temperature of ice box in a 3 star** (* * *) **fridge or deep freeze** Food will keep for three months (-15°C)

−30

Now answer these questions—

1 Between which temperatures do bacteria grow best?

2 Why is it important not to keep food lukewarm (20–40°C) for long periods?

3 Why will food not keep fresh for more than 2–3 days in the main part of a fridge?

4 Ultra heat treated milk keeps fresh for months, even without a fridge. Why is this?

Homework or further work on 'how to prevent food from going bad'

1 Look at this picture and then answer the questions below—

Dried foods—soup, rice, and cornflakes. Microbes cannot attack dried foods. Will keep for weeks in a sealed packet.

Canned foods. The food is heated to 120°C inside a sealed can. May keep for a year or more.

Milk—kept fresh by pasteurizing and keeping in a fridge.

'Quick dried foods' are dried in a vacuum at a low temperature (−18°C). Will keep for months.

Jam contains sugar Microbes cannot grow in sweet foods. Keeps for months or even years.

Bacon contains salt. Microbes cannot grow in salty food. Will keep for a week in a fridge.

Frozen food will keep for 3 months in a 3 star refrigerator or deep freeze.

Now copy out this table and fill it in by looking at ten foods kept in your kitchen at home—

Name of food	Why does it stay fresh?	How long will it keep?
1		
2		
3		
4		
5		
6		
7		
8		
9		
10		

2 Read pages 50–54 in **'Biology for the Individual, Book 7 War against disease',** and answer the questions you find there.

17 How to kill microbes in the home

One way to kill microbes is to use **disinfectants.**
Disinfectants contain chemicals which poison microbes.
But they must be used with care. Some people use them
without reading the instructions on the labels.

You can try using disinfectants which have been made up
according to the label, and then compare these to disinfectants
which have been diluted with too much water.

1 Collect six clean test tubes, and a test tube rack or tin, to hold them.
Make sure your tubes are clean—wash them and rinse them if necessary.

2 Find one of the screw cap bottles
containing a thin gravy. This
gravy contains millions of harmless
bacteria.

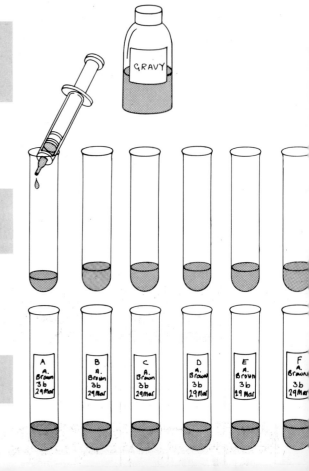

3 Use a clean syringe to transfer
3 cubic centimetres of the gravy to
each of your six tubes.

4 Label your tubes as shown—

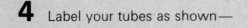

5 Find out which disinfectant is to be added to each tube. **Use a clean syringe each time—** wash out the syringe with water after filling each tube. Put 3 cubic centimetres of disinfectant in tubes A, B, C, D and E.

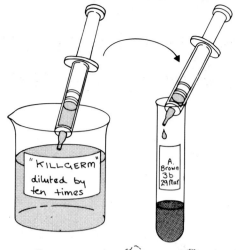

"KILLGERM" diluted by ten times

A. Brown 3b 29 Mar

6 **Put tap water in tube F.**

7 Plug each tube with cotton wool and leave in a warm place for several days.

A A. Brown 3b 29 Mar

B A. Brown 3b 29 Mar

C A. Brown 3b 29 Mar

D A. Brown 3b 29 Mar

E A. Brown 3b 29 Mar

F A. Brown 3b 29 Mar

8 Copy out this table of results—

		How do disinfectants affect bacteria?			
		Brand tested _____			
Tube	Strength of disinfectant added	At the start		5–7 days later	
		smell (fresh? decaying?)	appearance (clear? cloudy? coloured?)	smell (fresh? decaying?)	appearance (clear? cloudy? coloured?)
A					
B					
C					
D					
E					
F					

Write up the experiment in the usual way.

Why was it necessary to use tap water in the control tube (F)?

9 To be answered when your tubes are ready, seven days later—

The label on a well-known brand of mild disinfectant tells you— 'Add 1–2 tablespoonfuls to the water when you are having a bath'.

Would this amount be strong enough to kill any bacteria in the bath water?

18 How to kill bacteria inside you

1

Although disinfectants are a useful way
to kill bacteria around the house,
they are not much help when you are ill—

DISINFECTANT
NOT TO BE
TAKEN
INTERNALLY

No doctor will tell you to drink a bottle of disinfectant when you are ill. Why not?

2

When you are ill, it is important to find a way of killing any bacteria in your body, without killing you as well. Fortunately, there are several ways of doing this.

The most famous method was discovered in 1928, by the British doctor, **Alexander Fleming.** He was growing different kinds of bacteria on jelly plates at the time.

One day, he noticed a plate which had been spoilt—
a mould had grown near the edge of it.

Fleming's original plate.

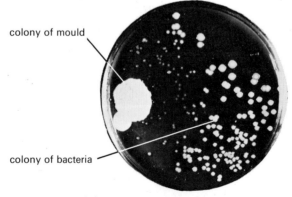

colony of mould

colony of bacteria

Fleming was about to throw it away, when he noticed something odd about the plate.

What was happening to the colonies of bacteria which were nearest to the mould?

3

Fleming noticed that the bacteria were not growing well near the mould. He asked himself—

'Is the mould making a chemical which is killing the bacteria?'

He then set up an experiment to test his idea. You can now carry out a similar experiment yourself, using a chemical extracted from Fleming's mould. This chemical is called **penicillin.**

You will be given discs of paper which have been soaked in different strengths of penicillin—

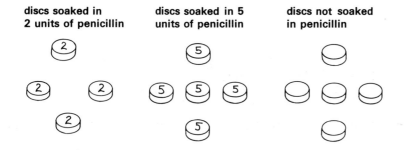

| discs soaked in 2 units of penicillin | discs soaked in 5 units of penicillin | discs not soaked in penicillin |

You can place each kind of disc on a jelly to which bacteria have already been added, and then leave the whole dish in a warm place for 24 hours.

Now follow these instructions—

1 Collect a bunsen burner, a pair of forceps (tweezers), and a dish of jelly which has already been treated with bacteria.

2 Label it on the base, with your name, class, and the date.

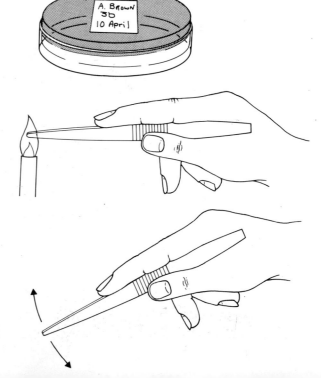

3 Heat your forceps in the tip of a flame for a few seconds.

4 Cool the forceps in the air for a few seconds.

5 Use the forceps to collect a
2 unit penicillin disc and place
this on the jelly.

Press the disc firmly into place.

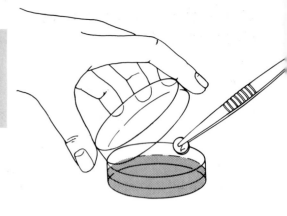

6 Heat the forceps again, cool them
in the air, and place a
5 unit disc on the jelly.

Press the disc firmly into place.

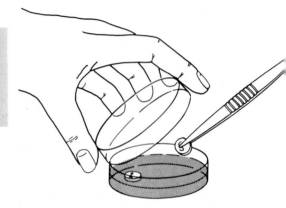

7 Heat the forceps again, cool
them in the air, and place
a **paper** disc on the jelly.

Press the disc firmly into place.

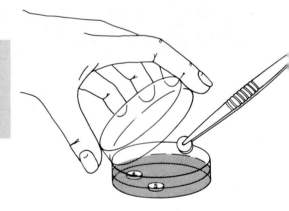

8 **Turn the dish over** and seal
it with sticky tape.
Leave upside down for 24 hours
at 37°C (or longer, at 20°C).

Now carry on to page 57 ⟶

A. BROWN
3b
10 April

Questions to be answered as soon as your dish is set up

1 Write up your experiment in the usual way.

2 Why was it necessary to flame the forceps each time they were used?

3 In your experiment, you used two paper discs soaked with penicillin, and one paper disc which had not been soaked with penicillin as a control. **Why was the control disc without penicillin used?**

4 **Can you ever remember being given penicillin by a doctor? If so, say when you were given penicillin, and why you were given it.**

5 Chemicals like penicillin are known as **antibiotics.** Antibiotics are used to kill bacteria in the body, but they will not kill other types of microbe.

modern antibiotics

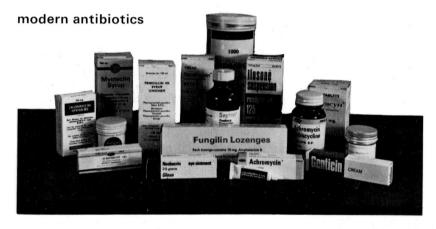

Have you ever been given antibiotics, apart from penicillin, by a doctor? Why were you given them? Were they helpful?

6 Antibiotics are not given to people suffering from diseases such as flu, mumps or chicken pox. **Why not?** (Hint—Which kind of microbe causes these diseases? See *Biology for the Individual, Book 7, pages 41–2.*)

Further work or homework

7 **Why is penicillin not as effective nowadays as it was in 1940, when it was first used?**

8 **Find out more about the discovery of penicillin, how it was first tested, and why two scientists once soaked their coats in it.**

9 **How is penicillin made nowadays on a large scale?**

19   Where do microbes come from?

1

If you boil up some grass in water, and leave it in a jar open to the air, microbes will appear in the water—

A drop of water from a freshly boiled grass and water mixture ($\times 30$).

A drop of water from the same mixture, three weeks later ($\times 30$).

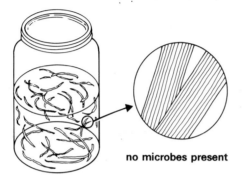

no microbes present

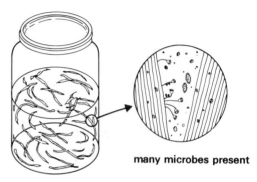

many microbes present

Where do you think these microbes have come from?

2

Most people nowadays know that the microbes probably came from the air, usually as tiny **spores.** When the spores land in a damp place, the microbe inside starts to grow.

But only a hundred years ago, many people would have said—
'The boiled grass has turned into microbes'.
The idea that living things could be formed from non-living things was also believed by ancient Greek scientists, such as **Aristotle** (384–322 B.C.).

Aristotle saw a dried up pond with no signs of life in it.

A week later, heavy rain fell, and eels appeared in the pond.

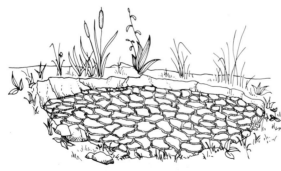

Aristotle suggested that the eels were formed by the action of rain on mud.

How do YOU think the eels got into the pond?

3

The Belgian scientist **van Helmont** (1577–1644), stated that mice could be 'made' by leaving wheat grains in a shed with an old shirt—

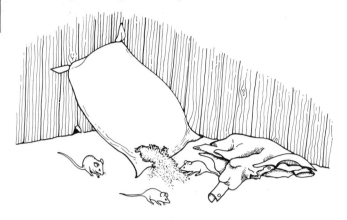

Where do you think the mice came from?

4

By the middle of the nineteenth century, most people realised that all of the larger animals and plants grew from eggs or seeds.

But many people still believed that microbes could grow from nothing inside a nourishing liquid. This idea was called **'spontaneous generation'**.

For example, the French scientist **Pouchet** carried out this experiment in 1860—

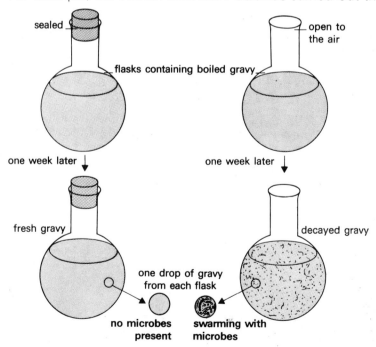

sealed

open to the air

flasks containing boiled gravy

one week later ↓   one week later ↓

fresh gravy   decayed gravy

one drop of gravy from each flask

no microbes present   swarming with microbes

Pouchet concluded that the microbes had been created by the action of oxygen on the soup.

Suggest a different explanation for his results.

5 The great French scientist **Louis Pasteur** decided to prove that Pouchet was wrong. He believed that the microbes in Pouchet's flasks had all come from the air, and that spontaneous generation could never occur.

To prove that he was right, he had to find a way to let air into the flasks without allowing microbes to enter as well.

So he set up this simple but ingenious experiment—

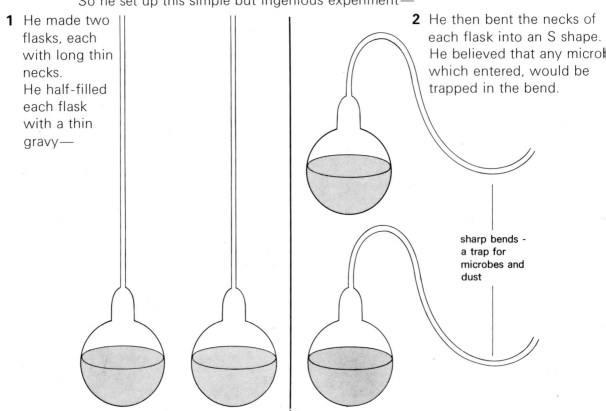

1 He made two flasks, each with long thin necks.
He half-filled each flask with a thin gravy—

2 He then bent the necks of each flask into an S shape. He believed that any microbe which entered, would be trapped in the bend.

sharp bends - a trap for microbes and dust

3 He boiled the gravy vigorously to kill all the microbes in both flasks—

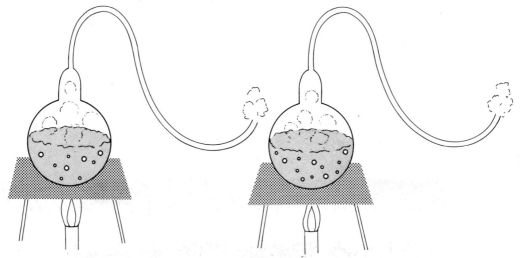

4 He then left both flasks undisturbed for several days.

Continued on page 61———

5 After a few days, he tilted one of the flasks so that the gravy ran along the curved tube.

6 He then tilted the flask back so that the gravy ran back into the flask, carrying with it any dust and microbes trapped in the bend.

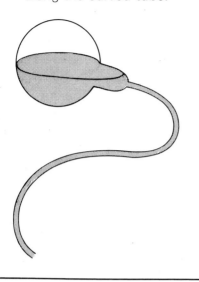

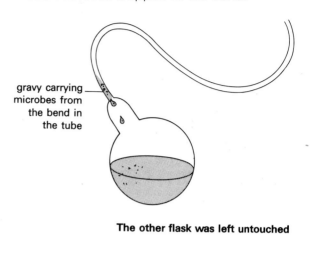

gravy carrying microbes from the bend in the tube

The other flask was left untouched

7 After a week, the gravy in the tilted flask was full of microbes, but the gravy in the untouched flask was still fresh.

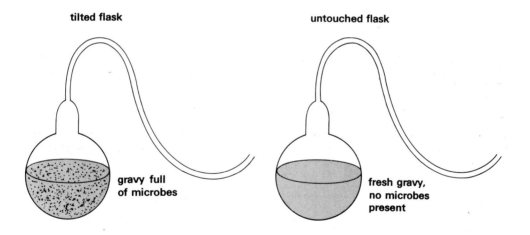

tilted flask

untouched flask

gravy full of microbes

fresh gravy, no microbes present

Pouchet claimed that gravy and air alone would produce microbes.
Pasteur claimed that gravy and air alone are not enough; he believed that microbes could only arise from existing microbes.

1 Who was proved correct by the experiment with the tilted flasks?

2 Give a reason for your answer to question 1.

6

Carrying out Pasteur's experiment for yourself

Pasteur carried out his experiment many times before he was certain of his results. Some of his flasks can still be seen to-day, over a hundred years later, at the Pasteur Institute in Paris.

The gravy is still there, just as fresh as when Pasteur first used it.

You may be able to carry out an experiment like his, for yourself—

To find out if microbes can be prevented from growing in gravy exposed to the air.

1 Collect—

two test tubes
a waterproof felt tip pen
a test tube rack
a small funnel, for pouring
two small pieces of cotton wool

two small pieces of cooking foil
two elastic bands
two pieces of glass tubing
   one curved and one straight,
   as shown below—

2 Set up your two tubes like this—

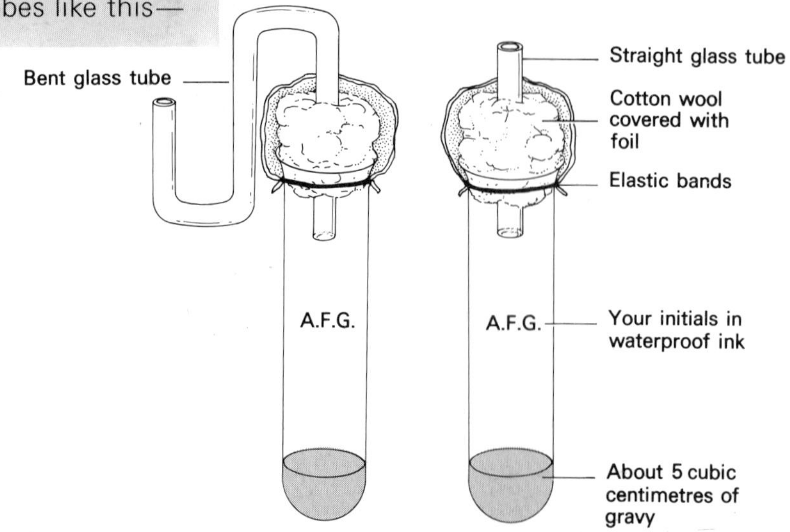

Bent glass tube

Straight glass tube

Cotton wool covered with foil

Elastic bands

A.F.G.

A.F.G.

Your initials in waterproof ink

About 5 cubic centimetres of gravy

3 Your two tubes should now be heated in a pressure cooker for 15 minutes, at a pressure of 10 newtons per square centimetre.

Meanwhile, write up your experiment in the usual way.

Further work or homework

During their long argument, Pasteur and Pouchet carried sealed flasks of gravy up glaciers high in the Alps and the Pyrenees.
Find out—(a) why they carried flasks up glaciers
         (b) what were the results of their experiments
         (c) how Pasteur explained his results
               (Reference— *'The microbe man'*, pages 44-47).